WITHDRAWN

Markham Public Library
Thornhill Community **Centre Branch**
7755 Bayview Avenue
Markham, ON L3T 4P1

Our Animal World

How Animals Sleep

by Emily C. Dawson

Amicus Readers are published by Amicus
P.O. Box 1329, Mankato, Minnesota 56002

Copyright © 2011 Amicus. International copyright reserved in all countries. No part of this book may be reproduced in any form without written permission from the publisher.

Printed in the United States of America at Corporate Graphics, North Mankato, Minnesota.

Library of Congress Cataloging-in-Publication Data
Dawson, Emily C.
  How animals sleep / by Emily C. Dawson.
     p. cm. – (Amicus readers. Our animal world)
  Summary: "Compares the different ways various animals sleep, including lying down, standing up, hanging, and more. Includes comprehension activity" – Provided by publisher.
  Includes index.
  ISBN 978-1-60753-014-5 (lib. bdg.)
  1. Animals–Sleep behavior–Juvenile literature. I. Title.
  QL755.3.D39 2011
  591.5'19–dc22
                                    2010007463

Series Editor   Rebecca Glaser
Series Designer   Kia Adams
Photo Researcher   Heather Dreisbach

Photo Credits

Alex Mares-Manton/Getty Images, 18; Corbis, 6, 21 (bottom); Corbis/Tranz, 8, 14, 16, 20 (both), 21 (top, middle); DAJ/Getty, cover; Dorling Kindersley/Getty Images, 10; infocusphotos.com/Alamy, 12; Janet Czekirda/123rf, 4–5; PixAchi/Shutterstock, 1

1224
42010

10 9 8 7 6 5 4 3 2 1

Table of Contents

Sleeping Animals	6
Picture Glossary	20
What Do You Remember?	22
Ideas for Parents and Teachers	23
Index and Web Sites	24

All animals need sleep. They need to rest because playing and hunting is a lot of work.

5

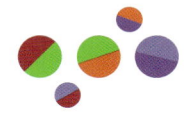

Cats sleep for about 12 hours every day. When cats wake up, their whiskers help them feel their way in the dark.

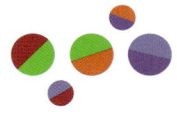

Dogs sleep less than cats. Sometimes dogs dream while they're sleeping.

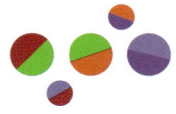

Hamsters sleep during the day. They curl up in a ball to keep warm.

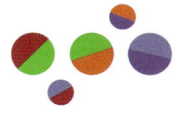

Flamingos sleep standing on one leg. They can stand on one leg for four hours.

claws

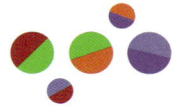

Owls sleep perched in a tree. Their claws lock on the branch so the owl doesn't fall.

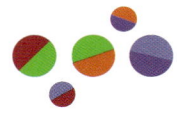

Bats sleep hanging upside down. Their talons grip a branch. Being upside down helps bats take off flying fast.

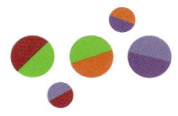

Humans need sleep, too.
How do you like to sleep?

Picture Glossary

claws
hard, curved nails on the foot of a bird

dream
to imagine events while you are asleep; Both humans and animals can dream

perched
sitting still on a branch in a tree

talons
a sharp claw of a bird or a bat

whiskers
the long, stiff hairs on a cat's face that it uses to feel its way around

21

What Do You Remember?

1. Trace this chart on a piece of paper.
2. For each animal, place an X in the box that tells how it sleeps.

Animal	Lying Down	Curled Up	Standing Up	Upside Down
cat				
dog				
hamster				
flamingo				
owl				
bat				
human				

If you don't remember, look back at the words and pictures in the book for the answers.

Ideas for Parents and Teachers

Our Animal World, an Amicus Readers Level 1 series, gives children fascinating facts about animals with lots of reading support. In each book, photo labels and a picture glossary reinforce new vocabulary. The activity page reinforces comprehension and critical thinking. Use the ideas listed below to help children get even more out of their reading experience.

Before Reading

- Talk about sleeping. Ask students where their pets sleep and how they sleep.
- Look at the cover photo together. Ask children what they think is happening.
- Look at the picture glossary words. Tell children to watch for them as they read the book.

Read the Book

- Read the book to the children or have them read independently. Remind them to look at the photos for clues if they need help understanding the words.
- Show children how to use the photo labels and picture glossary when they see an unfamiliar word.

After Reading

- Have children retell how animals in the book sleep, using the What Do You Remember? activity on page 22.
- See if anyone can think of other animals and how they sleep.
- Ask children if they always sleep the same way?

Index

bats 17
cats 7
claws 15
curl up 11
dogs 9
dreaming 9
flamingos 13
hamsters 11

humans 19
lie down 7, 9, 19
owls 15
standing up 13, 15
talons 17
upside down 17
whiskers 7

Web Sites

Science News for Kids: Why Cats Nap and Whales Snooze
http://www.sciencenewsforkids.org/articles/20080213/Feature1.asp

Neuroscience for Kids—Animal Sleep
http://faculty.washington.edu/chudler/chasleep.html

Sleep for Kids—Teaching Kids the Importance of Sleep
http://www.sleepforkids.org/

What Sleep Is and Why All Kids Need It
http://kidshealth.org/kid/stay_healthy/body/not_tired.html

Markham Public Library
Thornhill Community Centre Branch
7755 Bayview Avenue
Markham, ON L3T 4P1